Engineering for Inclusion

Held at the 2023 AIChE Annual Meeting

Orlando, Florida, USA
5-10 November 2023

ISBN: 978-1-7138-9318-9

Printed from e-media with permission by:

Curran Associates, Inc.
57 Morehouse Lane
Red Hook, NY 12571

Some format issues inherent in the e-media version may also appear in this print version.

Printed with permission by Curran Associates, Inc. (2024)

For permission requests, please contact AIChE
at the address below.

AIChE
120 Wall Street, FL 23
New York, NY 10005-4020

Phone: (800) 242-4363
Fax: (203) 775-5177

www.aiche.org

Additional copies of this publication are available from:

Curran Associates, Inc.
57 Morehouse Lane
Red Hook, NY 12571 USA
Phone: 845-758-0400
Fax: 845-758-2633
Email: curran@proceedings.com
Web: www.proceedings.com

TABLE OF CONTENTS

MAC EMINENT ENGINEERS AWARDS POSTER SESSION

MINORITY AFFAIRS COMMUNITY (MAC) AND YOUNG PROFESSIONALS (YP) PANEL DISCUSSION: REBRANDING YOURSELF FOR A NEW CAREER PATH

Assessing the Production and Cytocompatibility of *Plodia Interpunctella* silks as Polymeric Biomaterials

Jasmine McTyer, Bryce Shirk, Lauren Eccles, Isabella Torres-Duarte, Andrea Orozco, Paul Shirk, Whitney Stoppel

Silk is a versatile material that has been used for centuries to solve a variety of biomedical and environmental problems.Silk fibers are composed of many proteins, including silk fibroins and sericins, which contribute to the fiber structure and function.[1] Traditionally, engineers have relied on silk fibers and silk fibroin solutions derived from the domesticated silkworm, *Bombyx mori (B. mori)*, to develop new technologies.[1] Existing strategies for functionalizing and tuning mechanical properties of *B. mori* silk fibroin-based biomaterials have limitations. The large presence of highly crystalline domains makes applications in bioprinting and hydrogel systems difficult, especially under biologically relevant osmolarities and pHs.[2] However, there are a wide array of silk producing organisms that produce silk fibers with unique protein structure and functions that may overcome these challenges. These organisms include other moths in the Lepidopteran order and spiders. Currently, the utilization of alternative silk producing species has been limited due to low scale-up potential. However, laboratory rearing of *Plodia interpunctella (P. Interpunctella)*, a non-mulberry silk fiber producing insect, has shown promise as an alternative to *B. mori*. *P. interpunctella* is a Lepidopteran insect that produces silk fibers throughout its lifecycle to aid in survival and reproduction (**Figure 1A**).[3] To further explore the potential of *P. interpunctella* as a source of silk fibers for biomaterial production, we conducted laboratory rearing experiments to assess the influence of insect culture parameters on silk fiber production and insect growth. We manipulated temperature, population size, and food-to-caterpillar ratio to optimize silk fiber production while reducing energy inputs. Our results showed that higher temperatures accelerated insect growth and reduced life cycle length (**Figure 1B**). To maximize silk production relative to energy inputs, we varied the total number of larvae and the amount of food provided along with the temperature changes. We found that the number of larvae did not have a synergistic effect on silk fiber production, but variations in the ratio of larvae to grams of food, along with temperature, altered total fiber production (**Figure 1C**). We determined that culture of *P. interpunctella* in a 250 mL vessel at 24 °C, with 10 larvae per g of food, yielded the greatest advantage in terms of total silk fiber production. Additionally, we compared silk fiber yields to common *B. mori* silk fiber regeneration protocols and found that only 9 standard insect rearing boxes were needed to generate 5 grams of dry silk fibers in approximately 20 days, which amounts to ¾ of a standard insect incubator shelf. From these findings, we will continue to pursue *P. interpunctella* as an alternative to *B. mori* for silk fiber production due to its ease of laboratory rearing and control over fiber reproducibility, allowing for tighter control over environmental conditions.

Furthermore, we aim to establish a reference cytocompatibility response from raw *P. interpunctella* silk as we work to develop a silk solution regeneration protocol. Initial cytocompatibility was assessed using primary lung fibroblasts cells seeded onto raw *P. interpunctella* silk, raw *B. mori* silk, and regenerated silk films using *B. mori* silk fibroin as comparisons. Cytocompatibility was assessed by initial cell viability (24 hours) and total initial cell attachment. Cell-materials interactions were evaluated over 1 week in culture by monitoring cell proliferation rates, mitochondrial activity, and cell spreading on *P. interpunctella* materials vs. *B. mori* materials, demonstrating the utility of *P. interpunctella* silk fibers and regenerated silk biopolymer solutions for applications in tissue engineering and regenerative medicine.

Citations:

1. Rockwood DN, Preda RC, Yücel T, Wang X, Lovett ML, Kaplan DL. Materials fabrication from Bombyx mori silk fibroin. Nature Protocols. 2011;6(10):1612-31. doi: 10.1038/nprot.2011.379.

2. Wang Q, Han G, Yan S, Zhang Q. 3D Printing of Silk Fibroin for Biomedical Applications. Materials. 2019;12(3):504. PubMed PMID: doi:10.3390/ma12030504.

3. Silhacek DL, Miller GL. Growth and Development of the Indian Meal Moth, Plodia interpunctella (Lepidoptera: Phycitidae), Under Laboratory Mass-Rearing Conditions. Annals of the Entomological Society of America. 1972;65(5):1084-7. doi: 10.1093/aesa/65.5.1084.

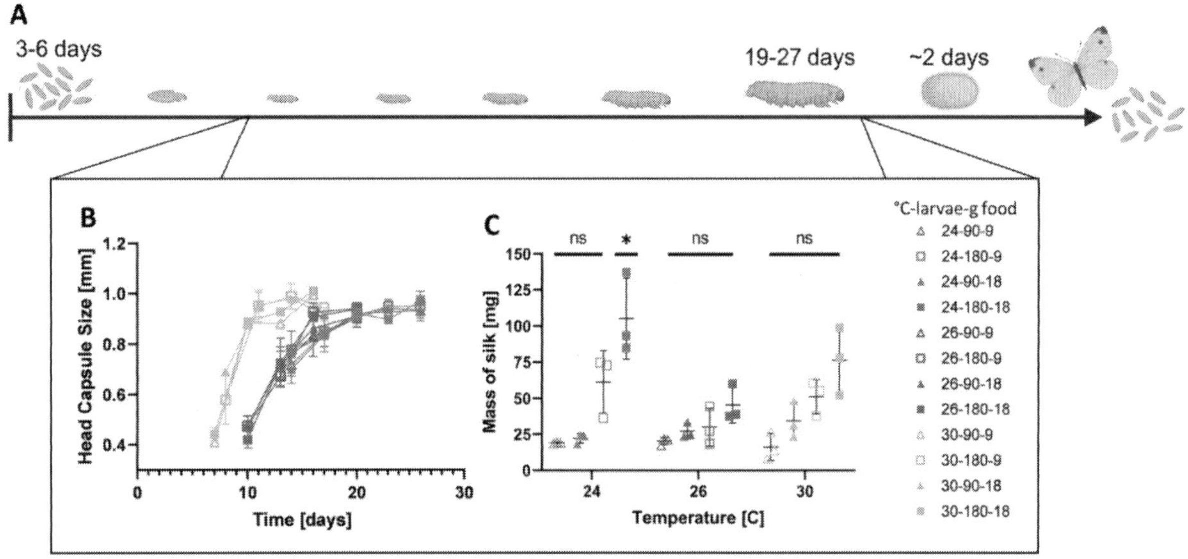

Figure 1: Culture conditions direct Plodia Interpunctella development. A) Life cycle of P. interpunctella. **B**) Head capsule measurements characterize how environmental conditions affect the growth cycle. **C**) Environmental impact on silk production. Larvae to food ratios are compared within one temperature parameter. *p<=0.05

Assessing the Feasibility of Coal As an Eco-Friendly Filler in Phenolic Resin Composites: A Study of Thermal and Structural Properties

Chiderah Chukwuka, Damilola Daramola, Sophia Almanza

Introduction: Due to the environmental concerns associated with using coal as the primary source of energy generation, there has been a recent surge of interest in exploring alternative utilization options for coal. The incorporation of coal particles as a filler in phenolic resin composites is a promising approach for composite development, owing to the similarity in the aromatic structure of both materials. This moiety similarity could promote increased interfacial connectivity and enhanced filler-composite compatibility.

In recent years, the demand for phenolic-based composite materials reinforced with fillers to reduce brittleness has risen due to their competitive performance and widespread use in the transportation, construction, and sports industries. Furthermore, fillers can provide an additional economic advantage by reducing cost; however, factors like filler effect on the curing process and the end use of the phenolic-based material must be considered. To fully exploit the potential of coal-based phenolic composites, it is essential to investigate the correlation between curing kinetics and network formation based on structural analysis. Understanding this correlation can provide insight into how different coal ranks can be leveraged to enhance certain mechanical and thermal properties of the composites, hereby promoting material advancement in the phenolic resin industry.

This work focuses on highlighting the effect of coal rank and content on the curing of phenolic resin using differential scanning calorimetry (DSC) and characterizing the cured composite samples using spectroscopic techniques.

Methods: DSC measurements were used to evaluate certain thermal properties like the peak temperature, heat of reaction, and activation energy. Two ranks of coal were selected due to functionality differences – anthracite and sub-bituminous coal type, and their content was utilized at varied filler weight percentages within the sample mix. The cured samples were characterized using FTIR and solid-state 13C NMR.

Results: The DSC analysis using a 40% filler weight content of both ranks of coal within the sample mix revealed an increased activation energy effect on the curing of phenolic resin with the sub-bituminous ranked coal (which could be attributed to the presence of a higher number of functional end groups) relative to the anthracite-ranked coal. Preliminary results from the FTIR analyses reveals a disappearance of aromatic isocyanate bonds (N=C=O) in the 40% coal-filled phenolic samples compared to the 100% phenolic, which suggests the possible interaction between the coal particles and the phenolic resin during cure. Results from the CNMR analysis was carried out to further elucidate the structural behavior of these cured samples and would be displayed during the poster presentation.

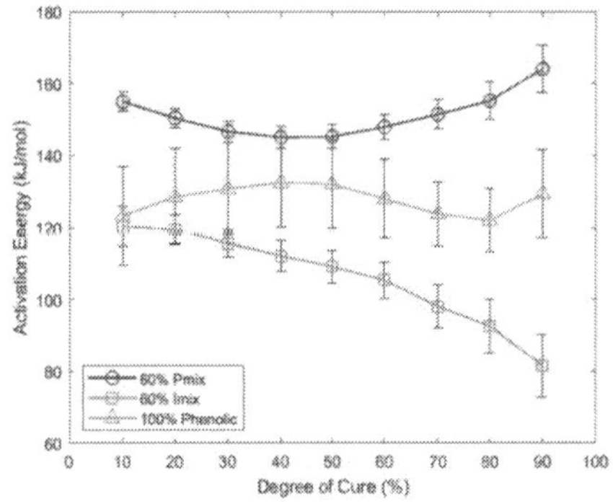

Activation energy values calculated utilizing the Friedman Method at various conversion stages and with samples containing 60% Pmix and 60% Imix.

60% Pmix - 60% phenolic resin/40% PRB subbituminous coal
60% Pmix - 60% phenolic resin/40% Itmann-Anthracite coal

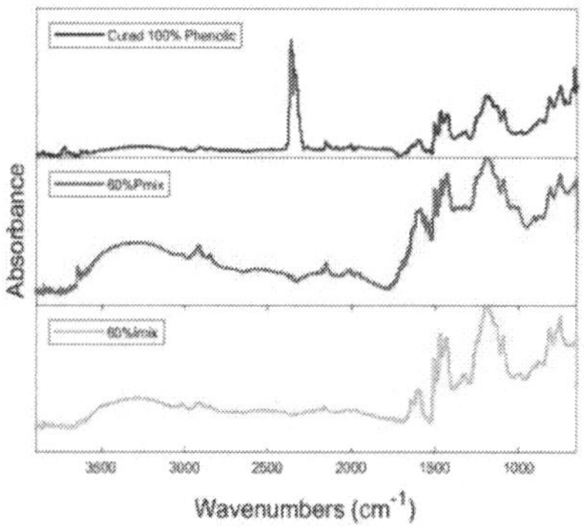

FTIR of cured phenolic resin vs 60% Pmix & 60% Imix @10 °C/min heating rate.

Exploring the Economic and Environmental Benefits of Solvent-Based Recycling Processes of Multi-Layer Plastic Films

Aurora Del Carmen Munguia Lopez, Dilara Goreke, Kevin Sanchez-Rivera, Horacio Aguirre-Villegas, Styliani Avraamidou, George Huber, Victor Zavala

Multi-layer plastic films are fascinating materials that are widely used for food packaging; these films are strategically designed to harness specific protection properties of different polymers (e.g., oxygen, moisture, and temperature barriers). These properties are essential to facilitate food storage and distribution in supply chains, as they enable long shelf lives, reduce refrigeration use, reduce food waste, and facilitate food accessibility. Unfortunately, multi-layer plastic films cannot be mechanically recycled using traditional approaches. As a result, large amounts of multi-layer films typically end up in landfills or are combusted. Solvent-based approaches have emerged as a promising alternative to separate and recycle the constituent polymers of multi-layer films. This process uses a series of solvent washes to dissolve and precipitate polymers in a targeted manner. Quantifying the economic and environmental benefits of solvent-based recycling processes is essential for their deployment and for enabling more circular economies.

This work presents a computational framework to conduct a detailed economic and environmental analysis of solvent-based recycling processes. The proposed framework integrates process simulation models, life cycle assessment methods, molecular simulations, and experimental data. Our analysis highlights critical components of the processes (specific unit operations and solvents) and of the multi-layer films (specific polymer layers) that have the highest economic and environmental impacts. I will discuss how our findings are currently guiding advances in process, product, and supply chain designs that are more sustainable.

Confirming the Extent and Mechanisms of the Immunosuppressive Enhancement of hMSCs Initiated By Col/Hep Layer By Layer Polyelectrolyte Coatings

Justin Putman, Roaa Hadi, Jorge Almodovar

Statement of Purpose: There has been an increasing need for highly potent human mesenchymal stromal cells (hMSCs) for the treatment of a number of chronic diseases. Though, for therapies, several millions of cells may be needed for each dose. A current method of reducing dosage and increasing therapeutic potential is to add soluble Interferon Gamma (IFN). Unfortunately, it has been shown that pretreatment with IFN suppresses proliferation of hMSCs. In order to mitigate this negative effect, cells were cultured on unique layer-by-layer (LbL) coatings, composed of collagen and heparin. A side effect found to accompany this approach is an enhancement of immunosuppressive properties. A main concern here is to identify the specific pathways by which the coatings enhance IFN activity. Doing so will allow further evolution of artificial extracellular matrices. First, the hypothesis that the LbL coatings negate the anti-proliferative effect of IFN and enhance IFN's activity by the activation of the Janus kinase (JAK) pathway, while also immobilizing IFN at proximity to the cell surface, creating a HEP+IFN complex that regulates IFN activity is to be tested.

Methods: The hMSCs were purchased from RoosterBio, sourced from several donors. Coatings (1) used 6.5 bilayers made up by sodium heparin from Celsius Laboratories, Inc., lyphosized type I collagen sponges from Integra Lifesciences Holdings Corp. and Poly(ethylene imine) (PEI) solution from Sigma-Aldrich, each in sodium acetate buffers at pH 5, 4 and 5 respectively. Multiple inhibitors are to be used. As of now, a JAK inhibitor, AG490 from Selleckchem and Sigma Aldrich has been used in the experiments. Upon seeding, the cells were incubated with either normal culture medium or inhibitor incorporated culture medium for 24 hours before being washed with PBS and incubated with either fresh culture medium or IFN incorporated culture medium(50ng/mL), for 3 or 6 days. Preliminary tests were done with PrestoBlue, on a 96 well plate to ascertain the minimum concentration needed for the drug to present as intended and if higher concentrations are toxic to the cells. Several concentrations were tested, ranging from 0.25 to 100μM. Upon settling on 1 μM, experiments began on 24 well plates, using a colorimetric assay for indoleamine-pyrrole 2,3- dioxygenase (IDO) with inhibited and LbL conditions to ascertain if and how much the coatings affect the amount of IDO secreted by the cells and thereby the effect on immunosuppressive properties. The assay works by first incorporating a standard solution to separated supernatant to permit IDO to convert L- tryptophan to N-formyl-kynurenine for 30 mins. This reaction is then stopped by 30% w/v trichloroacetic acid and heated at 58 °C. After cooling, the mixture is combined in a 50/50 ratio with 2% w/v p- dimethylaminobenzaldehyde in acetic acid into a 96 well plate for reading with a spectrometer at 490 nm to quantify the reaction of p-dimethylaminobenzaldehyde with kynurenine. The absorbance results are then normalized by cell count.

Results:

(Images Attached)

The PrestoBlue viability tests narrowed down the possible concentrations to 3 for use in subsequent experiments. The final PrestoBlue test shows that the concentrations under 1 micromolar don't have an extremely toxic effect on cells. And, preliminary experiments, and other studies (2) led to favoring concentrations of at least 1 μM for inhibition. Leading to a decision to use 1 μM for subsequent experiments involving the coatings. When cells were grown on TCP and No AG490, there was a large difference between the PG/Cell measured between the + IFN and - IFN conditions. When the layers were introduced, that gap was reduced immensely. The difference being 2.4-fold and 1.4-fold, respectively. When the AG490 was introduced the difference in the IDO secretion was decreased noticeably, with the ratio of magnitude being 58% without the coating and 84% with the coatings. The western Blot data, as well as proteomics experiments provided some further confirmation while also presenting more questions. Some of which will be addressed by proteomics data as well as other experiments to be included in Orlando in November.

Conclusions: Several preliminary conclusions were reached. Including that HEP/COL LbL coatings show an increase in IDO activity that is likely, to be in part, achieved through the JAK pathway. Also, it is of note to mention that the cell count for the conditions with the LbL was much higher, leading to a lower PG/Cell. Although, it must be mentioned that if cells died in the other conditions, as would be expected, given previous data with PrestoBlue, they still would have produced IDO that would be measured by the assay but more difficult to normalize. Still, more experiments are needed to further investigate and confirm the mechanisms and extent that the LbL coatings enhance immunosuppressive properties. Including protein analysis and tests with other inhibitors, which are currently in the works.

Results

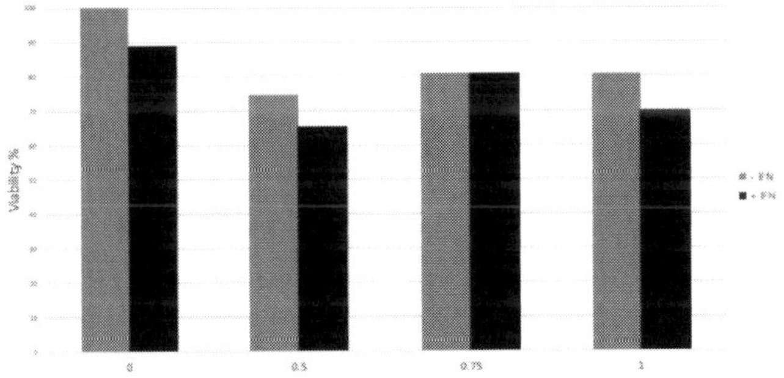

PRESTOBLUE- Cell viability of hMSCs with differing concentrations of AG490, an inhibitor of the JAK pathway, evaluate with Presto Blue cell viability assay. Here, one of the later iterations, where the concentrations were narrowed down to 3 from several (0-1000 micromolar), that show little toxicity to cells

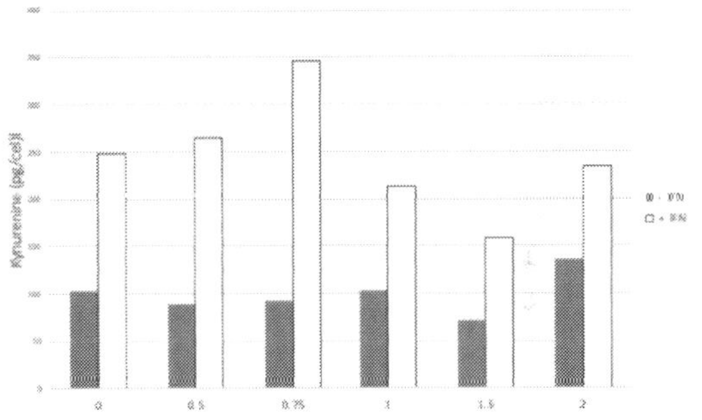

IDO EXPRESSION- immunomodulatory properties of hMSCs with differing concentrations of AG490, an inhibitor of the JAK pathway, evaluated via colorimetric assay. Here, the effective dose was evaluated to determine a minimum concentration of inhibitor necessary

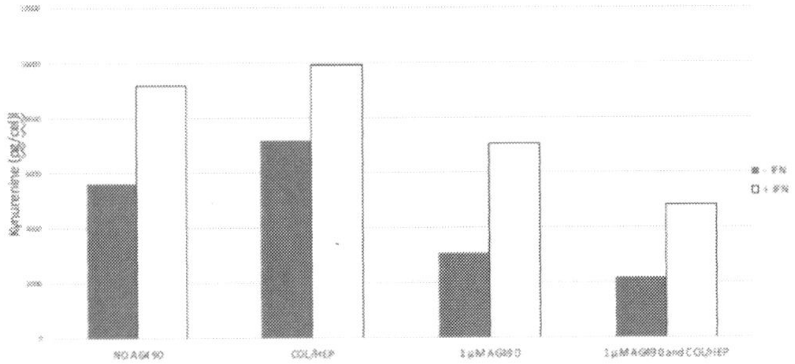

IDO EXPRESSION- Immunomodulatory Properties of hMSCs on TCP and 6.5 bilayers of Layer-by-Layer coatings, with and without AG490

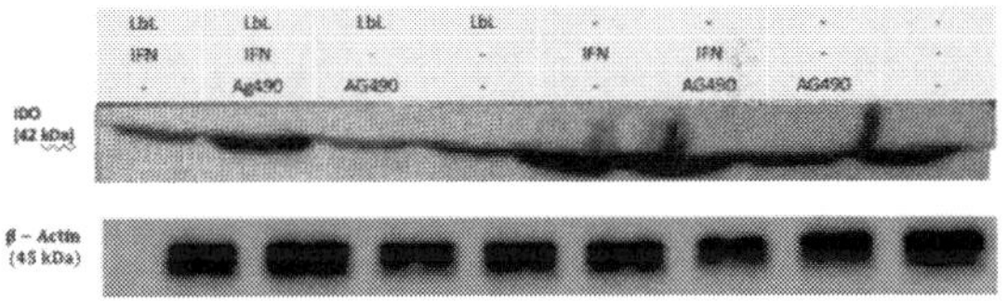

WESTERN BLOT- Protein Expression of IDO (Immunomodulatory Properties) and Beta-Actin for hMSCs on TCP and 6.5 bilayers of Layer-by-Layer coatings, with and without AG490

Unraveling the Enhancing Effect in Proliferation of Human Mesenchymal Stem Cells Cultured on Biopolymeric Coatings.

Roaa Hadi, Jorge Almodovar, Justin Putman, Luis Carlos Pinzon-Herrera, Shilpi Agrawal

Stem cell therapy is a type of tissue regeneration that-repairs damaged body cells by reducing inflammation and adjusting the immune system. To improve human stem cell capacity to modulate the immune system, a bioactive surface made of heterologous heparin and collagen was used. In this study, we examined the viability_and proliferation of hMSCs on different types of coatings with and without the_FGF-2 inhibitor drug AZD4547 at a concentration of 1 μM in the culture medium. Six different types of surfaces were analyzed, including a control surface (TCPS), a surface of 6 layers of HEP/COL, and 6.5 layers of HEP/COL with and without the_FGF-2 inhibitor drug AZD 4547. hMSCs were seeded on each surface of a 96-well plate, and after 0, 1, 2, and 3 days of culture.; For the viability of hMSCs, PrestoBlue's method was used. The cell culture medium was removed and replaced with 100 μL of PrestoBlue reagent per well and an incubation period of 3 hours to detect fluorescence. For the proliferation of hMSCs on COL/HEP coatings, detecting cells with EdU, 18 hours is the period of EdU exposure to hMSCs. Then, cells were fixed and permeabilized, EdU was detected, DNA was stained, and finally, images and data were analyzed. HEP/COL multilayers can increase hMSCs viability compared to the control surface. After 3 days, the results showed that the viability of hMSC was higher than that of TCPS, especially on surfaces with 6.5 heparin and collagen. Furthermore, when hMSCs are grown in COL/HEP coatings, their proliferation increases. Inhibition studies suggest that HEP/COL coatings may increase hMSC proliferation through activation of the FGF-2 pathway.

Enhancing Magnetic Particle Imaging (MPI) Performance: Probing Post-Synthesis Oxidation and Correlations to Nanoparticle Properties

Ambar Velazquez Albino, Aniela Nozka, Andrii Melnyk, Hayden Good, Carlos Rinaldi-Ramos

This study investigates the impact of post-synthesis oxidation on the performance of superparamagnetic iron oxide nanoparticles (SPIONs) in magnetic particle imaging (MPI), an emerging technology with applications in diagnostic imaging and theranostics. SPIONs synthesized from iron oleate were subjected to a post-synthesis oxidation treatment with a 1% Oxygen in Argon mixture. MPI performance, gauged via signal intensity and resolution using a MOMENTUM™ scanner, was correlated to the nanoparticles' physical and magnetic properties. Post-synthesis oxidation did not alter physical attributes like size and shape, but significantly enhanced magnetic properties. Saturation magnetization increased from 52% to 93% of the bulk value for magnetite, leading to better MPI performance in terms of signal intensity and resolution. However, the observed MPI performance did not fully align with predictions based on the ideal Langevin model, indicating the need for considering factors like relaxation and shape anisotropy. The findings underscore the potential of post-synthesis oxidation as a method to fine-tune magnetic properties of SPIONs and improve MPI performance, and the need for reproducible synthesis methods that afford finely tuned control of nanoparticle size, shape, and magnetic properties.

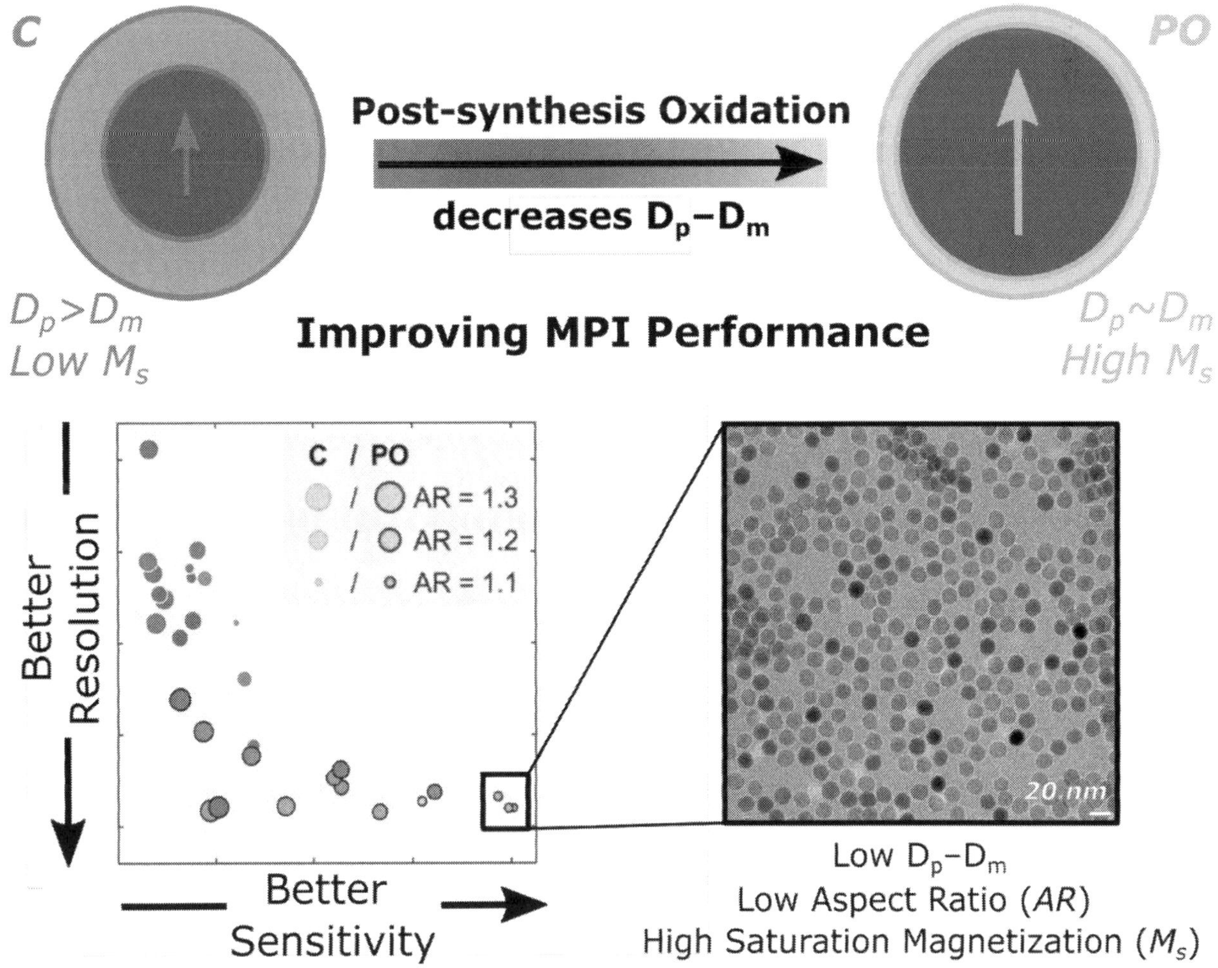

Electrified Lixiviation of Rare Earth Elements from Coal Mining Waste - MAC

Lawrence Ajayi, Damilola Daramola

Rare earth elements (REE) are a group of metals that are key drivers of the new-age economy, as these REEs are essential components of several high-tech technologies, including wind turbines, magnets, and video displays. With global increase in demand for green technology, about 12,000 metric tons of REE are consumed per year in the US and estimated to increase in the future[1]. However, more than 85% of the world's supply is from the Asia Pacific region[2], a risky proposition considering that any disruption in that region would have a significantly negative impact on the global supply chain of the aforementioned technologies. REEs has been identified to be present in coal-related resources[1], nevertheless, a key limitation to commercial extraction is the low concentrations of REE in these raw materials, typically in the ppb or ppm range[3]. Typical REE extraction methods, pyrometallurgy and hydrometallurgy, suffer from economies of scale associated with typical chemical processes, i.e., processes are only economical at large-scale centralized facilities[4]. Therefore, an extraction technique that is both modular (that can be applied at smaller decentralized resource locations due to flexible scalability) and highly selective (that can be tuned to target higher-value REE) would alleviate this REE supply concern by improving extraction economics.

Maes *et al.* explored the potential of enriching solvent with REE from Monazite using electrochemical extraction and found different leaching efficiencies for different REE with Nd extracted at a higher concentration than La[5]. The effectiveness of this electrochemical driving force and the selectivity of REE at different potential from an alternative less-concentrated REE coal pre-combustion waste is unknown. This research's objective is to enhance the lixiviation (i.e., extraction of a soluble material) of rare earth elements from solid coal mining waste residues (50:50 coal to ash composition) through the addition of an electrochemical stimulus to chemical extraction design variables. Specifically, applying an electrical stimulus below that for hydrogen generation introduces a voltage gradient that induces diffusion of the higher concentration ionic species (typically the adsorbents for the REE) to the electrode surface, and therefore increases the availability of the REE at the electrode. As an added advantage, these electrified techniques can be coupled with renewable energy sources such as solar or wind, thus mitigating the need for fossil energy consumption[6].

In this study, to be presented at the annual meeting, the effect of process variables such as temperature, cathodic potential, agitation speed and solvent type on extracted REE concentration in the liquid phase and selectivity of each REE will be reported. Furthermore, preliminary process economics will be reported via specific energy consumed in the process.

References:

(1) Vass, C. R.; Noble, A.; Ziemkiewicz, P. F. The Occurrence and Concentration of Rare Earth Elements in Acid Mine Drainage and Treatment By-Products: Part 1—Initial Survey of the Northern Appalachian Coal Basin. *Min. Metall. Explor.* **2019**, *36* (5), 903–916. https://doi.org/10.1007/s42461-019-0097-z.

(2) Hower, J. C.; Granite, E. J.; Mayfield, D. B.; Lewis, A. S.; Finkelman, R. B. Notes on Contributions to the Science of Rare Earth Element Enrichment in Coal and Coal Combustion Byproducts. *Minerals* **2016**, *6* (2), 32. https://doi.org/10.3390/min6020032.

(3) Lin, R.; Howard, B. H.; Roth, E. A.; Bank, T. L.; Granite, E. J.; Soong, Y. Enrichment of Rare Earth Elements from Coal and Coal By-Products by Physical Separations. *Fuel* **2017**, *200*, 506–520. https://doi.org/10.1016/j.fuel.2017.03.096.

(4) Zhang, W.; Rezaee, M.; Bhagavatula, A.; Li, Y.; Groppo, J.; Honaker, R. A Review of the Occurrence and Promising Recovery Methods of Rare Earth Elements from Coal and Coal By-Products. *Int. J. Coal Prep. Util.* **2015**, *35* (6), 295–330. https://doi.org/10.1080/19392699.2015.1033097.

(5) Maes, S.; Zhuang, W.-Q.; Rabaey, K.; Alvarez-Cohen, L.; Hennebel, T. Concomitant Leaching and Electrochemical Extraction of Rare Earth Elements from Monazite. *Environ. Sci. Technol.* **2017**, *51* (3), 1654–1661. https://doi.org/10.1021/acs.est.6b03675.

(6) Sedlak, D.; Mauter, M.; Macknick, J.; Stokes-Draut, J.; Fiske, P.; Agarwal, D.; Borch, T.; Breckenridge, R.; Cath, T.; Chellam, S.; Childress, A.; Dionysiou, D.; Giammar, D.; Hoek, E.; Jiang, S.; Katz, L.; Kim, J.; Kostecki, R.; McCutcheon, J.; Polsky, Y.; Stoll, Z.; Xu, P. *National Alliance for Water Innovation (NAWI) Master Technology Roadmap*; NREL/TP-6A20-80705, 1818076, MainId:77489; 2021; p NREL/TP-6A20-80705, 1818076, MainId:77489. https://doi.org/10.2172/1818076.

Extraction and Characterization of Cellulose Nanocrystals from Corn Husk Waste and Its Application in Bioplastic Preparation

Sergio Mayta Paucara, Maria Quintana Caceda, Ronny G. Huamani Palomino, Bryan M. Córdova

Maize is the most cultivated cereal in the world. In fact, according to data of the Food and Agricultural Organization of the United Nations, global production reached up to 1.16 billion tons in 2020. In Peru, according to data from the Minister of Agriculture (MIDAGRI, 2021), more than 414 thousand tons of corn were harvested in 2020, mainly in the coastal valleys of Lima and Ica regions.

In this context, cereals production has also given rise to considerable amounts of waste, which are the residues generated by the harvesting and processing of the feedstock. In particular maize wastes include stalks, chaffs, steams, silks, and husks; out of these components' stalks constitute more than half of the stover biomass while the husk accounts for 10% of the total mass. Furthermore, these wastes barely degrade on the soil surface, making their disposal still a challenge, especially in developing countries where the lack of efficient and environmentally friendly techniques ultimately generates environmental threats.

In this sense, open air burning is one of the main disposal strategies for corn waste in Peru, however, this practice has been widely questioned by the scientific community as it constitutes one major source of gaseous and particulate emissions to the atmosphere. In fact, both greenhouse gasses (carbon dioxide (CO2), carbon monoxide (CO), ie.) and volatile organic compounds (ammonia (NH3), sulfur dioxide (SO2), ie.) would get released to the surroundings.

On the other hand, scientists have shown a particular interest in the valorization of different lignocellulosic biomass over the last decade, as it constitutes the largest source of renewable organic material on Earth. In particular, cellulose has received remarkable attention considering its outstanding mechanical properties (tensile strength between 5 and 7 GPa along the chain direction) and its versatility in the synthesis of added-value derivates.

Moreover, micro and nano forms of cellulose are quite relevant because of their unique properties conferred at such a small scale. Cellulose nanocrystals (NCC) are rod-like particles with typical widths and lengths in the range of 5–70 nm and several micrometers, respectively. NCC exhibits low density (~1.46 g/cm3), an outstanding mechanical performance (bending strength estimated at 10 GPa and elastic modulus ranging from 10 and 100 to 200 Pa, respectively), and a high surface area/aspect ratio of 10 up to 70.

Under this context, this research focuses on the valorization of Peruvian corn wastes into cellulose nanocrystals using both mechanical and chemical treatments combined.

The feedstock was collected from local markets in Lima, Peru, and after washing it several times, the sample was dried, milled, and sieved under 850 μm. Afterward, dried corn husk powder was treated with 4% (w/v) sodium hydroxide twice. Then, peracetic acid bleaching was conducted according to a previous methodology. This cellulose sample was characterized by instrumental techniques and gravimetric measurements. Finally, acid hydrolysis was conducted with concentrated sulfuric acid at 45°C followed by centrifugation, dialysis and low-frequency ultrasonication. The nanocrystalline cellulose suspension obtained was used to prepare films by simple casting method.

The cellulose was extracted with a 79.4% yield while its purity was estimated at 70.3%. On the other hand, in order to identify characteristic functional groups of spectral bands, characterization by FT-IR spectroscopy was carried out, and characteristic bands of cellulose were found at 1426 cm-1, 1325 cm-1, 1163 cm-1, and 1110 cm-1. Likewise, the decrease in signals at 1251 cm-1 and 1736 cm-1 indicates that hemicellulose and lignin were removed after the chemical treatments.

At the same time, x-ray diffractograms showed peaks at 22.5° and 34°, characteristics of the crystal polymorphs of cellulose, while the crystallinity index was estimated in 54 by the Segal equation. SEM images showed a change on the structure of the material from irregular and rough to a highly fibrillated network arrangement.

The average hydrodynamic diameter of the isolated CNC was determined by dynamic light scattering (DLS). The nanocrystals dimension was found to be 268 nm, however, its important to highlight that the hydrodynamic diameter represents the dimension of an equivalent sphere for different types of particles (individual size, aggregates, and agglomerates) and thus may not represent the actual size of the particle, in fact, CNC is known for having a high length: width ratio (aspect ratio). Therefore, aiming to obtain a better understanding of the morphology of the crystals, atomic force microscopy was performed into diluted suspensions, it was found that the nanocrystals had a rod-like shape, with an average aspect ratio of 35.

Furthermore, for comparison, commercial nano cellulose was also prepared and the dimension was found to be 218 nm.

It is also important to mention that an interesting effect caused by low-frequency ultrasonication was noticed. As the sonication time increased both the viscosity and the particle size of the nanocrystalline suspension decreased. This effect might be a consequence of the rearranging of the particles caused by cavitation. Previous studies have found that ultrasound induces a decrease in the aspect ratio of suspension while promoting the individualization of particles. Thus, further research studying these effects is still needed.

CNCs films were cast by pouring the suspension into a petri dish and drying overnight. These films were highly transparent and brittle. Their resistance to both alkaline and acidic medium was evaluated by measuring the weight loss over the days and comparing it to polyethylene films.

Regarding the water absorption capacity, the films were able to absorb their weight up to 3 times, approximately. It was noted that the bioplastics obtained had good resistance to acids (sulfuric acid) and bases (sodium hydroxide) up to concentrations of 25% and 40%, respectively. After 5 days there was no loss in the mass of the CNCs films. When these films are compared with those commercially obtained (polyethylene), it was observed that the latter lose 20% of their mass in the presence of 40% alkali, while the former was intact.

This evidence could respond to the high stability of the crystalline structure and the strong bonding between molecules, making it suitable for future applications in the medical and food industries, where these properties are highly appreciated.

Demonstration of a Batch Electrochemical System for Phosphorus Recovery from a Real Municipal Wastewater Recycle Stream - MAC Session

Sana Heydarian, Syed Asad Abbas, Lawrence Ajayi, Jason Trembly, Damilola Daramola

Phosphorus is an essential nutrient for both animals and plants, but it has an adverse effect on the ecosystems of water streams due to eutrophication [1]. Phosphorus recovery from wastewater streams is an essential part of wastewater treatment plants to curb the destruction of ecosystems in downstream water bodies. The idea of recirculation of phosphorus from wastewater treatment plants to grain fields will not only provide a continuous fertilizer supply but it will also minimize the scarcity issues associated with phosphorus in the world of growing phosphorus demand [2]. Sustainable process of the phosphorus recovery in the form of fertilizer from wastewater treatment plants requires optimal reactor design, high recovery rates, and minimum energy consumption [3]. Electrochemical systems have an additional advantage of being utilized using renewable energy resources. In this study we have established a rational design for the sustainable removal of phosphorus using an electrochemical system in the form of struvite or brushite, which are magnesium-based and calcium-based phosphates, respectively [4]. A complete range of experiments were performed for the investigation of phosphorus recovery efficiency and the specific energy consumption during recovery. Chronoamperometry, Cyclic voltammetry and Impedance analysis were performed for a real wastewater sample to investigate the reaction mechanism, and energy consumptions. Inductively Coupled Plasma (ICP) was utilized to quantify the recovery rates in the liquid streams, and Scanning Electron Microscopy (SEM), X-ray Diffraction (XRD) and Raman analysis were performed for quantitative and qualitative analysis of recovered solid products. These results will be presented at the upcoming conference.

References

[1] Y. Wang *et al.*, "Sulfur and iron cycles promoted nitrogen and phosphorus removal in electrochemically assisted vertical flow constructed wetland treating wastewater treatment plant effluent with high S/N ratio," *Water Research*, vol. 151, pp. 20–30, Mar. 2019, doi: 10.1016/j.watres.2018.12.005.

[2] P. Devos, A. Filali, P. Grau, and S. Gillot, "Sidestream characteristics in water resource recovery facilities: A critical review," *Water Research*, vol. 232, p. 119620, Apr. 2023, doi: 10.1016/j.watres.2023.119620.

[3] G. P. Pindine, J. P. Trembly, and D. A. Daramola, "Equilibrium-Based Temperature-Dependent Economic Analysis of the Recovery of Phosphorus from Different Wastewater Streams via Chemical Precipitation," *ACS EST Water*, vol. 1, no. 11, pp. 2318–2326, Nov. 2021, doi: 10.1021/acsestwater.1c00166.

[4] Z. Belarbi, D. A. Daramola, and J. P. Trembly, "Bench-Scale Demonstration and Thermodynamic Simulations of Electrochemical Nutrient Reduction in Wastewater via Recovery as Struvite," *J. Electrochem. Soc.*, vol. 167, no. 15, p. 155524, Nov. 2020, doi: 10.1149/1945-7111/abc58f.

Hydrophobic Collapse of Long Alkane Chains in Water and Water/Ethanol Mixtures

Rahul Kumar, David Abadie, Nicole Korinetz, Henry Ashbaugh

Protein folding is of tremendous importance in biology, but the mechanism of folding is still not well-understood. Various non-biopolymers in "good" solvents have been studied extensively to mimic such dramatic collapse of chain conformations, and the behavior has been explained in terms of hydrogen bonding and preferential adsorption. But little attention has been paid to the role of hydrophobic effect in such cases. To unravel the importance of hydrophobicity, here, we choose three long alkane chains (C20, C30, C40) as our model systems. A single chain in water, and water/ethanol mixtures is simulated using all-atom replica exchange molecular dynamics technique at temperatures (*T*s) between 274 K and 366 K. In pure water, at all *T*s, we see a two-state-like behavior in the radius of gyrations (R_g) of the chain conformations with the chain arranging into either a coil (larger R_g) or a globule (smaller R_g). The relative fractions of two populations, f_C and, f_G are quantified using 2D probability densities of chain conformations constructed with R_g and number of native contacts as two variables. We find that f_G systematically increases with T for the chosen alkanes. Further, C20 remains mostly extended and C40 mostly collapsed at all *T*s while C30 shows a crossover between f_C and f_G at an intermediate T. We estimate the free energy of such transitions based on f_C and f_G and explain the collapse behavior in terms of enthalpic and entropic components of the free energy. On addition of small amount of ethanol in water, we find that chain conformations are more collapsed compared to the pure water case, showing a signature of cononsolvency. At this composition, ethanol is found to be preferentially adsorbed near the chain, aligned with literature. In the pure ethanol case, however, chains swell and adopt more extended conformations. These observations could provide insights to the role of hydrophobicity in polymer collapse and protein folding.

MAC Poster: Mathematical Modeling of Drug Release from Bi-Layered Drug Delivery Systems in the Eye

Eduardo A. Chacin Ruiz, Jabia M. Chowdhury, Katelyn E. Swindle-Reilly, Ashlee N. Ford Versypt

Wet age-related macular degeneration (wet AMD) is a chronic disease that involves the formation of new blood vessels in the retina due to overexpression of vascular endothelial growth factor (VEGF). It progressively affects the central retina and can cause vision loss if left untreated. Wet AMD is the leading cause of visual impairment in the Western World and the third leading cause worldwide. There is no cure for this disease; however, there are therapies based on anti-VEGF drugs. The most common administration route of anti-VEGF therapeutics is intravitreal injections, but due to the short half-life of the anti-VEGF drugs, monthly intravitreal injections are needed. These injections are expensive, painful, and can result in poor patient compliance. Furthermore, other complications, such as retinal detachment, can also arise due to repeated intravitreal injections. To address this problem, several drug delivery systems (DDS) have been proposed in the literature to extend anti-VEGF release in the eye and reduce the frequency of injections. In this work, we detail our efforts at mathematical modeling of drug release from chitosan-polycaprolactone (PCL) DDSs with different geometries to help optimize their design and improve wet AMD treatments.

First, we model the prototype chitosan-PCL spherical microparticles consisting of an inner chitosan core and an outer PCL shell [1] with the drug dispersed within the core. Fick's second law is used to model the unsteady-state drug release into phosphate buffer saline. We assume constant diffusion coefficients in each polymeric layer, no convection or reaction within the DDSs, negligible erosion and swelling effects, and perfect sink conditions at the interface between the DDS and the release medium. We consider the existence of a partition coefficient at the interface between chitosan and PCL. For the drug-loaded chitosan-PCL microspheres, we solve the diffusion equation in spherical coordinates numerically using a finite differences algorithm in MATLAB. Nonlinear least squares curve fitting is used to determine the drug diffusion coefficients in both materials, the initial burst release, the partition coefficient between chitosan and PCL, and the critical time at which burst effects are not significant anymore. We then reconstruct the same system in COMSOL Multiphysics, which uses a finite element algorithm, to compare the capabilities of the MATLAB and COMSOL optimization routines for curve fitting in advance of extending to more complicated geometries. We use this process to verify our numerical techniques for integrating the amount of drug remaining in each layer of the microspheres and for calculating the objective function for minimizing the sum of squared errors between the experimental data and the model predictions for cumulative drug release over time [1].

We then build a COMSOL model with the dimensions of the chitosan-PCL microcapsule, which is a bi-layered cylinder with a hollow core for drug loading. Transport of dilute species was the only physics enabled in COMSOL as we first focused on the diffusive transport of drug from the DDS into the *in vitro* medium. The boundary condition is set to be a no flow condition on the top and bottom of the DDS, simulating impermeable boundaries, consistent with the fabrication techniques of sealing the end caps of the device. The DDS is assumed to be under perfect sink conditions, therefore the drug concentration on the outside is kept at zero. Fick's second law is used to model the unsteady-state drug release from the cylinder. Since our fabrication technique of salt-leaching the outer PCL layer allows us to modulate the DDS porosity, we are fitting the experimental data [2] for the various salt leaching conditions to correlate the impacts of the initial porosity on the subsequent drug release trajectories. Following a similar procedure as in the microparticles, we used nonlinear least squares curve fitting to estimate the drug diffusion coefficients in both materials, initial burst release, partition coefficient between polymers, and the critical time.

In parallel, we are developing a 3D human eye geometry model which involves relevant tissues' dimensions and boundary conditions. This model can be used to predict drug transport within the eye. The model considers both convective and diffusive transport. The vitreous humor of the eye is regarded as a porous media, and therefore Darcy's law is applied to model the convective flow within the vitreous. Additionally, both anterior and posterior segment elimination pathways are considered.

Our MATLAB and COMSOL models accurately simulated the cumulative drug release behavior from the microspheres for 160 days compared to in vitro experimental data. Furthermore, we observed similar behavior for radial drug concentration profiles within microspheres. For the cylindrical device, we observed large deviations in the initial 50 days, with more accurate predictions after that, implying that other drug-release mechanisms, like erosion, need to be considered for the initial phase. In addition, we used the 3D eye model to study the pharmacokinetic behavior of drug after intravitreal injection in two locations: near the limbus, and at a proposed new target location.

The models can help optimize the design of bi-layered DDSs to improve wet AMD treatment and provide insights into the mechanisms involved in the drug release from these DDSs in the eye. In the future, our refined model will be able to calculate cumulative drug release profile, therapeutic release time, and DDS depletion time from input conditions of drug loading, DDS geometry, and initial porosity. Furthermore, we will consider the effects of porosity over the drug release from the DDS and use the COMSOL optimization module to estimate the relevant parameters. System-level integration will be achieved by combining our model for drug release from DDSs with the human eye geometry model, so that we can simultaneously simulate drug release and drug distribution within the eye. The use of this holistic model alongside with new in vitro and in vivo experimental data will also allow us to optimize the DDS design by tuning its thickness and porosity, so that its drug release is extended as much as possible while keeping the drug concentration in the eye within a therapeutic range.

References:

1. Jiang, P., Jacobs, K. M., Ohr, M. P. & Swindle-Reilly, K. E. Chitosan-Polycaprolactone Core-Shell Microparticles for Sustained Delivery of Bevacizumab. *Mol Pharm* **17**, 2570-2584, doi:10.1021/acs.molpharmaceut.0c00260 (2020).
2. Jiang, P., Chaparro, F. J., Cuddington, C. T., Palmer, A. F., Ohr, M. P., Lannutti, J. J., Swindle-Reilly, K. E. Injectable biodegradable bi-layered capsule for sustained delivery of bevacizumab in treating wet age-related macular degeneration. *J Controlled Release* **320**, 442-456, doi:10.1016/j.jconrel.2020.01.036 (2020).

Acknowledgment: This work was supported by the National Institutes of Health grant R35GM133763 and the University at Buffalo.

Biorefrineries for a Sustainable Plastic Value Chain

Efrain Rodriguez-Ocasio, Mark Blenner, Laura R. Jarboe

Plastic pollution is a major environmental challenge because most plastics are not biodegradable and strategies to prevent accumulation in ecosystems have been insufficient. Biorefineries for plastic waste upcycling offer a solution by diversifying the portfolio of products made from plastic waste and enabling a circular economy to disrupt the plastics value chain. To make this possible, the biorefinery would require the breakdown of plastic polymers and the subsequent bioconversion by microbial cell factories. This research proposes a hybrid thermal and biochemical process to upcycle polyethylene (PE), the most common type of plastic. The process combines thermal oxo-degradation (TOD) to transform PE into viable carbon feedstock for the nonconventional yeast *Candida maltosa*. This yeast was selected for microbial cell factory development from a screening of 20 species against model compounds representative of the feedstock and it grew from PE feedstock at rates comparable to glucose. The proof-of-concept results were remarkable, considering that the PE carbon feedstock is insoluble in water and solid. Adaptive Laboratory Evolution increased the growth rate by >200% and provided insights into the uptake mechanisms used by *C. maltosa* to overcome the mass transfer limitations in liquid culture. Characterizing the parent and evolved phenotypes identified canal structures formed for hydrophobic feedstock utilization, an increase in the secretion of biosurfactants and their ability to solubilize hydrophobic molecules instead of emulsifying them. These findings may inform rational strain engineering strategies to optimize industrial strains for plastic biorefineries and refineries using other hydrophobic feedstocks. In the meantime, this work is entering the metabolic engineering phase with the goal of engineering *C. maltosa* to produce value-added chemicals from PE waste.

Tri-Component Microneedle Patch for Enhanced Veterinary Meloxicam Delivery: PVA, Collagen, and Chitosan Synergies

Jorge Almodovar

Statement of purpose: Pain management for farm animal has become an important issue for different organizations in the agricultural sector, including farmers, veterinary associations organizations like the American Veterinary Medical Association (AVMA), animal welfare organizations, and governmental bodies (Frida L., Animal. 2018; 8:4-16). They are working collaboratively to address pain management in farm animals to ensure their well-being, uphold ethical standards, and meet consumer expectations encouraging the use of pain relief during routine management practices. However, few pain medications exist commercially that are Food and Drug Administration (FDA)-approved for use in these animals. Meloxicam is a promising NSAID alternative, with a relatively long half-life of 28 h. Currently, meloxicam is FDA-approved and routinely prescribed for pain mitigation in other veterinary species (i.e., dogs and cats) (Dominique V., Animal. 2018; 8:35). However, the administration of meloxicam to farm animals typically involves injections and oral intake using suspensions or tablets, which is not suitable for livestock treatment. This study suggests a more effective method for delivering drugs to farm animals is by applying polymeric microneedle patches transdermally. Microneedle patches composed of polyvinyl alcohol (PVA), type I collagen (COL), chitosan (CHI) and meloxicam (MEL) were evaluated regarding their physical, chemical, and mechanical properties, and the microneedle insertion abilities in pig's ear cadaver skin for the transdermal delivery of meloxicam as a pain management approach in veterinary applications.

Methods: The physical properties of the microneedle patches, including morphology, topography, and distribution were examined before and after ethylene oxide sterilization using Scanning Electron Microscopy (SEM) and a laser microscope 3D & profile measurement. A cross sectional imaging of the bilayer conformation of the microneedle patch was analyzed and visualized with a laser scanning confocal microscope (LCSM). Fourier transform infrared spectroscopy (FTIR) was used to analyze the chemical composition of microneedle patches with and without drug before and after the sterilization process. The microneedle patches were tested against the capacity to penetrate the pig's ear cadaver skin inner and outer surface at 5 s and 24 h. Finally, mechanical tests were conducted using a tribometer to assess the penetration depth of drug-loaded microneedles into the cadaver skin of pig's ear at different loads. These tests were also used to determine the deformation behavior of the microneedles compressed against the flat end of a stainless-steel pin (compression tests).

Results: The microneedle patch consisted of 225 microneedles with 600 μm of height and 300 μm of base uniformly organized on an area of 8 mm × 8 mm. The penetration of the microneedles is through the epidermis and dermis where the microneedle degradation and, consequently, the drug release begins (Fig. 1A-B). The optical cross-section patch side view validated the division and the difference between the two layers of the patch (Fig. 1C). SEM characterization demonstrated the presence of microneedles uniformly organized on the patch surface and preserve their morphological properties after the sterilization process. (Fig. 1D). FTIR characterization confirmed that microneedle patches with and without drug preserve the chemical composition of the PVA, COL, and CHI polymers and meloxicam drug after the sterilization process using ethylene oxide gas (Fig. 1E). In-vitro microneedle insertion study demonstrated that microneedle patches were able to penetrate pig's ear cadaver skin (Fig. 1F). Full penetration of the microneedles into the skin can be obtained by applying approximately 3 N, and microneedles have the capability of returning to their original shape after compression testing (Fig. 1G).

Conclusions: In summary, this study demonstrates that PVA-COL-CHI-MEL patches presented an organized distribution and homogeneous dimension of microneedles. The morphological properties and chemical composition after the sterilization process were preserved. The resultant microneedle patches were successful in penetrating the surface skin in pig's cadaver ear. Finally, microneedles have the capability of returning to their original shape after compression testing. This work demonstrated that this system of biodegradable microneedle patch will help to extend the use of the FDA-approved meloxicam in the pharmaceutical field how pain management medication for livestock animals as a suitable way to manage pain in farms animals after routine procedure.

Figure 1. PVA-COL-CHI-MEL microneedle patch. **A)** Macroscopy picture of the microneedle patch. **B)** Schematic illustrations of transdermal delivery of meloxicam using PVA-COL-CHI-MEL microneedle patches. **C)** Optical microscopy image of the microneedle patch cross-section area. **D)** SEM images of non-sterile (i) and sterile (iii) microneedle patch. **E)** FTIR spectrums of non-sterile and sterile **F)** In-vitro imaging of microneedle insertion in pig's ear cadaver skin **G)** Mechanical property of microneedle patch.

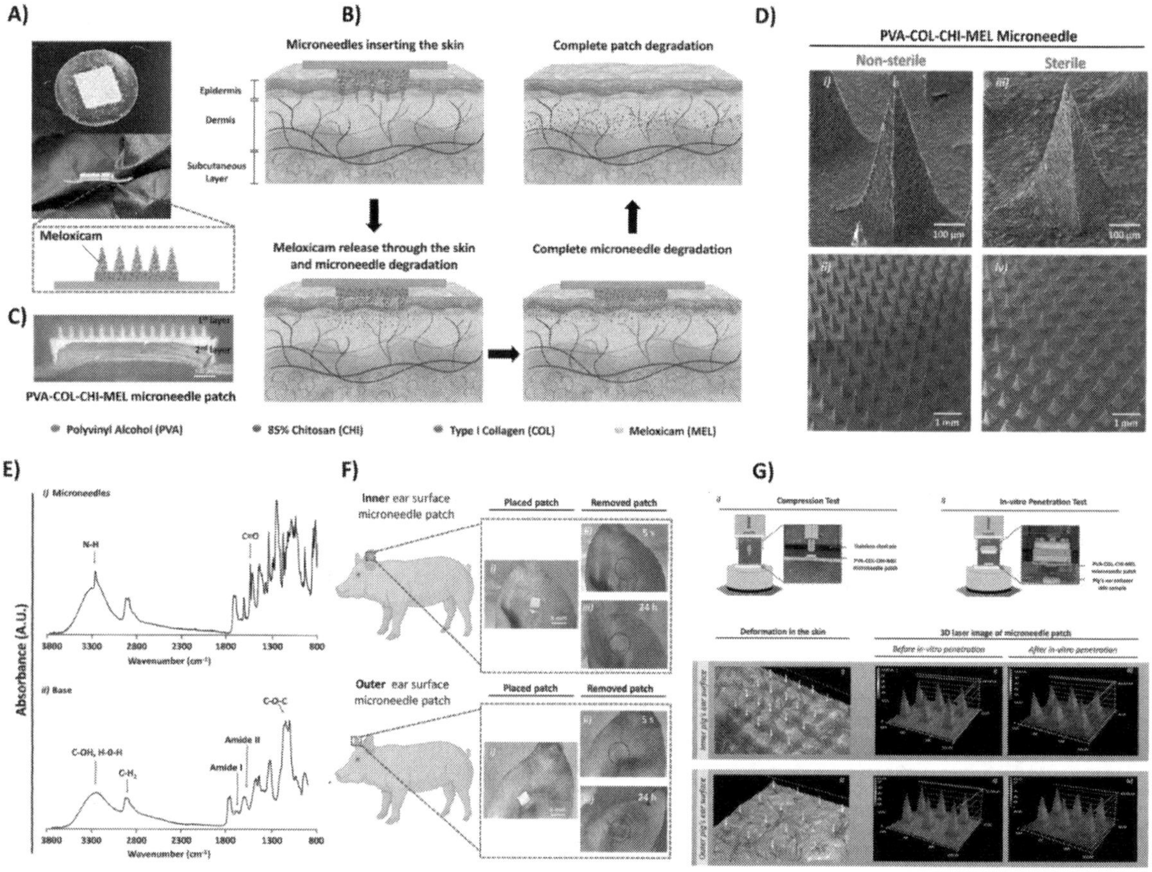

Dielectrophoresis-Based Breast Cancer Study: Characterization and Separation of Peripheral Blood Mononuclear Cells from Pymt and WT Mouse Model

Raphael Oladokun, Christopher Smith, Timothy Eubank, Soumya Srivastava

In 2021, the US is expected to see 284,200 new cases of breast cancer, making it the top cancer in women, with an estimated 43,600 deaths. Breast cancer is a systemic disease that affects multiple body systems and can be detected in circulating immune cells before tumor cells enter circulation due to their abundance in the body. However, early detection and proper treatment can enhance breast cancer outcomes. To realize our long-term goal, we are probing the dielectric properties of peripheral blood mononuclear cells (PBMCs) from peripheral blood sources of MMTV-PyMT mice at 14+ weeks (stage IV) using a microfluidic platform.

The central hypothesis of this research is that changes triggered in the subcellular components, such as the cytoskeleton, lipid bilayer membrane, cytoplasm, focal adhesion proteins, and extracellular matrix (ECM) at the onset of carcinoma regulate dielectric properties (conductivity, σ, and permittivity, ε), thereby affecting the bioelectric signals that aid in the detection of breast cancer. This hypothesis is developed based on our preliminary data demonstrating unique dielectric properties of PBMCs obtained from WT and PyMT tumor-bearing mice. The results obtained from our preliminary analysis identify the bioelectric signals that regulate human adenocarcinoma cells.

We hypothesize that isolated peripheral blood mononuclear immune cells (PBMCs) are altered in a body that has cancer compared to those PBMCs in a healthy body. This alteration is evident in the mouse model we are using, which is a powerful tool for analyzing the mechanisms of human breast cancer induction. Although the specific mechanisms behind these PBMC changes are not being investigated here, we hypothesize that these alterations can be detected using dielectrophoresis, a dielectric characterization technique. Changes in the electrophysiological properties of the plasma membrane of cancer cells are among the key features that support the use of dielectrophoresis to manipulate and carry out electrokinetic separations of normal or healthy cells from cancer cells and vice versa. This is because they tend to behave differently under a non-uniform electric field.

Our results present the dielectric properties of murine PyMT and WT PBMCs, which exhibited unique cellular behavior, and numerical simulations to validate these results. We conclude that these unique characteristics can be used to discriminate between cancer and non-cancer cells. This novel tool is label-free, rapid (~2 min.), and low-cost cell sorting technology that detects and separates early and late stages of breast cancer. This work will lead to preclinical development and future clinical trials of the developed detection platform. The long-term goal is to bring this analysis into the clinic to be able to non-invasively determine breast cancer at the earliest stages without the false positive and false negative rates of standard screening methods like mammography.

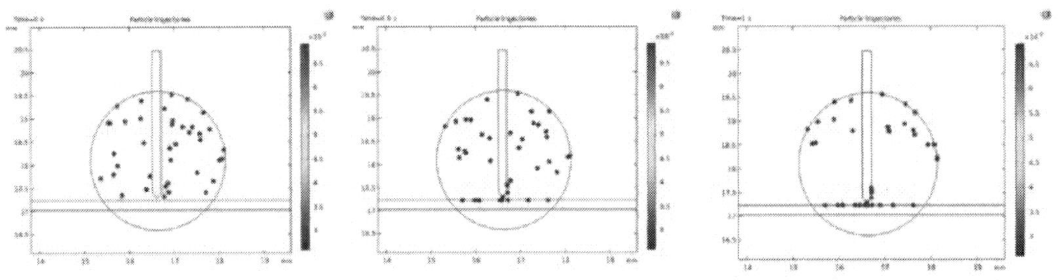

(A) DEP response of PyMT and WT-PBMCs at 0.02 S/m, 250 kHz, and 8 Vpp. (●) represents PyMT PBMCs, which experience pDEP, and (●) represents WT PBMCs, which experience nDEP. The distinct responses of the cells under the same electric field suggest a favorable separation frequency region for the two cell types.

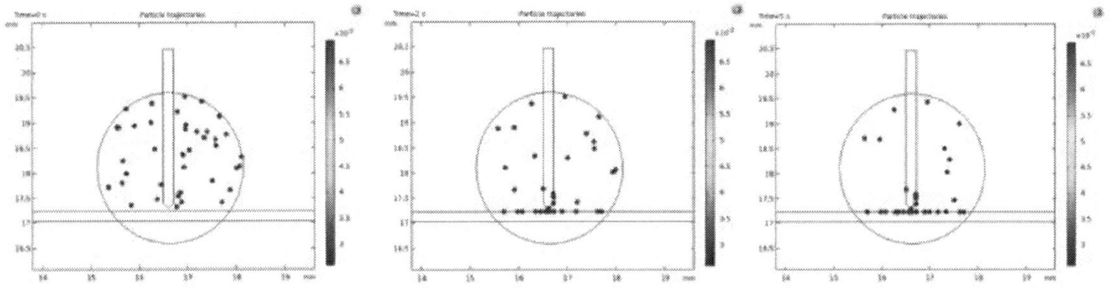

(B) Positive DEP response of PyMT and WT PBMCs at 0.02 S/m, 300 kHz, and 8 Vpp. This response validates the experimental results obtained during the characterization experiment of the study, as reported []. (●) represents PyMT PBMCs, and (●) represents WT PBMCs.

Figure1 Simulation of DEP characterization of PyMT-PBMC and WT-PBMC

106a Panel Discussion: Rebranding Yourself for a New Career Path

MaryKathryn Lee

This panel will discuss ways you can (re)brand yourself and your resume for your next job. This panel will feature members from the AIChE Board of Directors who have changed their career path and Young Professionals who have used many resources to broaden their job responsibilities.

Tell us your goals and hear from folks who were or are on a similar journey while learning about resources that AIChE can offer you.

AUTHOR INDEX